CIRCLE'S TRUE PI VALUE EQUALS THE SQUARE ROOT OF TEN

$$\pi = \sqrt{10} = 3.16227766016838$$

or

$$\pi = 3.1623$$

Albert Vitales Cruz, PhD

Professional Engineer (retired)
United States Citizen
United States Army Veteran

Copyright © 2023 by Albert Vitales Cruz, PhD.

Library of Congress Control Number:		2023912101
ISBN:	Hardcover	979-8-3694-0241-2
	Softcover	979-8-3694-0240-5
	eBook	979-8-3694-0239-9

All rights reserved. No part of this book may be reproduced or transmitted in any form or by any means, electronic or mechanical, including photocopying, recording, or by any information storage and retrieval system, without permission in writing from the copyright owner.

Any people depicted in stock imagery provided by Getty Images are models, and such images are being used for illustrative purposes only. Certain stock imagery © Getty Images.

Print information available on the last page.

Rev. date: 08/22/2023

To order additional copies of this book, contact:
Xlibris
844-714-8691
www.Xlibris.com
Orders@Xlibris.com
853624

DEDICATION

Dedicated to: My parents Juan Perez Cruz and Leonora Vitales Cruz;

My deceased wife Linaflor Manantan Cruz; and my siblings Ernesto Vitales Cruz, Josefina Cruz Calayag, Gerardo Vitales Cruz, Erminda Cruz Mora, Corazon Cruz Millena, and Salvador Vitales Cruz

PREFACE

The traditional Pi value of 3.1415 is found in many formulae in trigonometry and geometry, especially those concerning circles, ellipses, and spheres. Geometry is one of the oldest branches of mathematics. It is concerned with properties of space that are related to distance, shape, size, and relative position of figures. Until the 19th century, geometry was almost exclusively devoted to the fundamental concepts of Euclidean geometry, which includes the notions of point, line, plane, distance angle surface, and curve. Later in the 19th century, the scope of geometry has been greatly expanded, and the field has been split into many subfields that were known as *combinatorial geometry*. Other scholars used various mathematics equations to shape the circumscribed polygon into an inscribed circle by using an ever-increasing number of polygon sides to add more decimal places into the traditional Pi value of 3.1415. Indeed the Pi has been known for almost 4000 years, but even if the number of minutes that elapsed since then the calculated Pi to that number of many decimal places added to the 3.1415 is still only approximating its actual value.

ABSTRACT

The author of this book has discovered an innovative method of determining the True value of Pi ($\pi = \sqrt{10} = 3.1623$ or 3.16227766016838). This new Pi value is derived from the geometric relationships among the circle's components with the use of the Circle Theorem and Pythagorean Theorem. *Figure 1* contains an inscribed circle in the square consisting of gridlines equally spaced into one-fourth of the side of the square or the diameter of the inscribed circle. The resulting precise Pi value is validated with the use of the Polygon Area formula, Binomial Theorem, and Quadratic Equation. This contemporary approach to finding the true Pi value reputes the traditional method of finding the Pi value which represents the ratio between the circumference and the diameter. For the past four centuries, many mathematicians have attempted to find the precise Pi value. It began with measuring the circumference and the diameter of a circle and dividing the former by the latter. The Pi calculation began during the era of Archimedes of Syracuse circa 287–212 before the Christian era (BCE). Archimedes one of the greatest mathematicians of the ancient world introduced the approximate value of Pi as 3.14 (between 3-1/7 and 3-10/17 bound). Since then, humans have been trying to add more digits to the two-decimal placed Pi to find a more precise approximation of Pi with varying degrees of success. The amateur mathematician William Shanks, for example, calculated Pi by hand to 707 figures in 1873 and died believing so, but decades later it was discovered he had made a mistake at the 528th decimal place.

Jan de Gier, a professor of mathematics and statistics at the University of Melbourne, says being able to approximate Pi with some precision is important because the mathematical constant has many different practical applications. "Knowing Pi to some approximation is incredibly important because it appears everywhere, from the general relativity of Einstein to corrections in the GPS to all sorts of engineering problems involving electronics," de Gier says. In maths, Pi pops up everywhere. "You can't escape it," says David Harvey, an associate professor at the University of New South Wales. In 1897, the State of Indiana in the US almost did away with fussy strings of decimals altogether. The State's bill is almost enshrined in the law that π = 3.2 because the trigonometric method to shape a square into a circle is a mathematical impossibility.

The ubiquity of π makes it one of the most widely recognized mathematical constants used in elementary mathematics and scientific research. Several books devoted to π have been published, and record-setting calculations of the digits of π often result in news headlines. It is an impressive and time-consuming feat but this may not be necessary. The race to introduce more decimals into the erroneous Pi equals 3.1415 may be useless because this new approach to finding the true Pi value is so precise.

CONTENTS

Dedication v
Preface vii
Abstract ix
Chapter 1 Introduction 1
Chapter 2 Research Study 7
Chapter 3 Literature Review 23
Chapter 4 Results 35
Chapter 5 Summary 37
Chapter 6 Recommendation 39
Reference 41
Appendix *A* Glossary (Terms definitions) 45
Appendix *B* Notary Public Document 53

CHAPTER 1
INTRODUCTION

The author of this book Albert V. Cruz precisely proves that the value of Pi (π) equals the square root of ten ((π = 3.16227766016838) or π = 3.1623 rounded to four decimal places, which is the true ratio between a circumference and a diameter of the circle. To prove this true Pi value, the author utilizes the Circle Theorem and the Pythagorean Theorem, as well as the Binonial and Quadratic Equations. The use of these theorems and equations validates the author's prior logical statements that the circle's Arc ($\pi D/4$) equals the sum of the isosceles triangle's equal sides or the Hypotenuse of a Base (full Chord) and the Altitude (half Chord) formed by a 90-degree central angle.

This new Pi value reputes the old traditional approximation of Pi (π = 3.1415). The ancient Babylonians calculated the area of a circle by taking 3 times the square of its radius, which gave a value of Pi = 3 (i.e., $3r^2 = \pi r^2$). One Babylonian tablet (ca. 1900–1680 BCE) indicates a value of 3.125 for Pi, which is a closer approximation. In the Egyptian Rhind Papyrus (ca.1650 BCE), there is evidence that the Egyptians calculated the area of a circle by a formula that gave the approximate value of 3.1605 for Pi. The ancient cultures mentioned above found their approximations by actual measurement of the circumference and the diameter of a circle in which the ratio between the former and the latter represents a Pi.

The Ancient Greek mathematician Archimedes (circa 170 BCE) discovered an effective method for approximating the value of Pi (π). With the use of a polygon, he inscribed a regular hexagon in a circle and then circumscribed another regular hexagon in the same circle. He obtained a rough approximation of Pi (π) by dividing the Archimedes approximated the area of a circle by using the Pythagorean Theorem to find the areas of two regular polygons. The polygon is inscribed within the circle and the polygon within which the circle was circumscribed. Since the actual area of the circle lies between the areas of the inscribed and circumscribed polygons, the areas of the polygons gave the approximate area of the circle. Archimedes knew that he had not found the true value of Pi. He acknowledged that his Pi was only an approximation within those limits. Archimedes also used a 96-side polygon, which helped him find the closest approximation of Pi (π) as the original straight sides began to shape into a circle. The Archimedes number of 3.14 has remained one of the most prevalent approximations of Pi (π) ever since.

A similar approach was used by Zu Chongzhi who could not have been familiar with Archimedes' method—but because his book has been lost, little is known of his work.

Zu calculated the value of the ratio of the circumference of a circle to its diameter to be 3.1459 (355/113). To compute this accuracy for Pi, he must have started with an inscribed regular 24,576-sided polygon and performed lengthy calculations involving hundreds of square roots carried out to 9 decimal places. Mathematicians began using the Greek letter π in the 1700s. Introduced by William Jones in 1706, the use of the symbol was popularized by Euler, who adopted it in 1737. An 18th-century French mathematician named Georges Buffon introduced a method to calculate Pi based on probability.

Since ancient Babylonian times, humans have been trying to introduce more digits to the approximated constant with varying degrees of success. To find a precise value of Pi, in the mid-20th century, Swiss researchers spent 108 days calculating traditional Pi to a record of 62.8tn digits. Using a computer, their approximation beat the previous world record of 50tn decimal places and was calculated 3.5 times as quickly. The amateur mathematician William Shanks, for example, calculated Pi by hand to 707 figures in 1873 and died believing so, but decades later it was discovered he had made a mistake at the 528th decimal place.

Though not accurate, the traditional Pi value of 3.1415 is found in many formulae in trigonometry and geometry, especially those concerning circles, ellipses, and spheres. Geometry is one of the oldest branches of mathematics. It is concerned with properties of space that are related to distance, shape, size, and relative position of figures. Until the 19th century, geometry was almost exclusively devoted to the fundamental concepts of Euclidean geometry, which includes the notions of point, line, plane, distance, angle, surface, and curve. Later in the 19th century, the scope of geometry has been greatly expanded, and the field has been split into many subfields that were known as *combinatorial geometry*. Other scholars used various mathematical equations to shape the circumscribed polygon into an inscribed circle by using an ever-increasing number of polygon sides to add more decimal places to the original Pi value of 3.1415. Indeed Pi has been known for almost 4000 years, but even if the number of minutes that elapsed since then the calculated Pi to that number of many decimal places is still only approximating its actual value.

There is a need to find a precise or true value of the Pi of the circle but not by adding more digits to the erroneous traditional Pi value. Instead of the traditional approach to shaping the polygon into a circle, this book introduces a new innovative approach contained in the following paragraphs that utilize the Circle Theorem and the Pythagorean Theorem to establish the relationship between the arc of a circle and the resulting chord, which in turn, yield a precise Pi value. The method of proof uses geometric and math equations and proves that the equivalent straightened arc of a circle is the hypotenuse of the base chord and the altitude one-half chord that is formed by a 90-degree central angle. The arc (πD/4) is equated to the hypotenuse of a right triangle formed by the rectangle's diagonals and pairs of isosceles triangles as well.

CHAPTER 2
— RESEARCH STUDY —

The following geometric math derives the precise Pi value. *Figure 1* below consists of the inscribed circle in the square that is divided into 4 equal sections on both sides. It also includes 4 equal chords three equal arcs, and diagonals, all of which are pertinent lines for calculating the precise Pi (π) value.

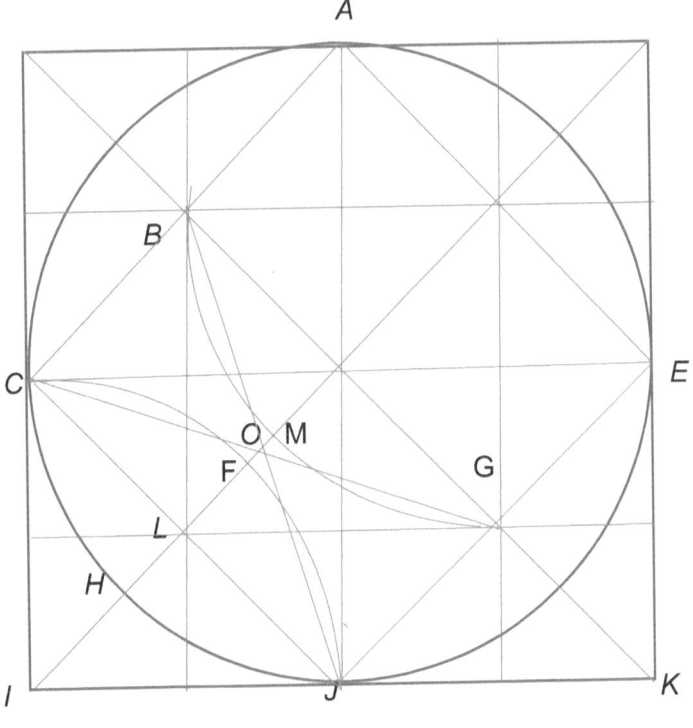

Figure 1

Figure 1 Technical Description

$CE = AJ$ is the Diameter D of the Circle or the side of circumscribed Square

$IJ = CI = FI$ is the Radius R of a Circle   Radius is one-half of the diameter.

$AC = CJ = EJ = AE$ is the Chord of the Circle   Lines whose endpoints both lie on Arcs.

Polygon ACEJ is inscribed square in a circle

$BC = GJ = \frac{1}{2}AC = \frac{1}{2}EJ$ is one-half of the Chord.	Point G lies in the middle of EJ; Point B lies in the middle of AC.
AC is Perpendicular to CJ	Adjacent equal chords are Perpendicular.
BC is Perpendicular to CJ	BC is a part of Chords.
CG is the Hypotenuse of Base CJ and Altitude GJ.	Pythagorean Theorem applies to the right triangle CGJ.
BJ is the Hypotenuse of Base CJ and Altitude BC	Pythagorean Theorem applies to right triangle BCJ.

CG and BJ are equal pairs of diagonals of rectangle $BCGJ$

Arc CFJ = Arc CHJ = BMG = $\pi D/4$ Arc with 90-degree central angle equals ¼ Circumference. Points F and M lie in the middle of Arcs CFJ and BMG. Points F and M lie along the diagonal of the square which equally divides the Arcs. The large diagonal from Point I to the opposite corner bisects Arcs CHJ and BMG. Point O is in the intersection of the diagonals of rectangle CBGJ.

Point D is not used in *Figure 1* as it is reserved for the Diameter D of this article.

Analyses of the circle's components

It is important to note that in *Figure 1*, all terminals or endpoints from A through M lie in the intersections of the gridlines D/4. These conditions are essential in establishing the relationships among the components of a circle. Each of the terminals of the Diameter D, the Radius ½ D, the equal Chords AC, AE, EJ, and JC, a swell as the equal Arcs CHJ, CFJ, and BOG lie at the intersection of the gridlines. The Diameter (CE) is the hypotenuse of the Base AC and Altitude AE. The Chord CJ is the hypotenuse of the Radii CI and IJ. Likewise, EJ is the hypotenuse of EK and JK. Other components of a circle such as the Tangent and the Secant lines are not analyzed in this article due to their irrelevancy in deriving a Pi. Besides, the right triangle and its hypotenuse are also necessary to determine the ratio between the one-fourth of the Circumference ($\pi D/4$) and the Diameter D, which yields the Pi (π). The key to determining the precise value of Pi defends where the straight-line equivalent of Arc ($\pi D/4$) lies among the other components of the circle. Hence, the Pythagorean Theorem rules in establishing the relationships among the circle's components.

Arc and Chord Analyses

Arc CHJ = CFJ = $(90°/360°)\pi D$ = $\pi D/4$ The Arc $\pi D/4$ is also ¼ of Circumference.

Line CG is the hypotenuse of the right triangle CGJ consisting of Chord CJ and half Chord GJ.

$(CJ)^2 = (D/2)^2 + (D/2)^2$ Pythagorean Theorem: The sum of the squares of the Altitude and Base of the right triangle equals the square of the Hypotenuse.

$(CJ)^2 = (1/4 + 1/4) D^2$ Factor out the common multipliers D^2

$(CJ)^2 = (1/2) D^2$ Fractions ¼ and ¼ added

$CJ = D/\sqrt{2}$ — Squareroot on both sides of the equation. (Base of right triangle CGJ)

$GJ = \tfrac{1}{2} CJ = D/2\sqrt{2}$ — One-Half Chord (Altitude of right triangle CGJ)

Arc *CFJ* and Arc *BMG* Analyses

If the mid-point of the Arcs at Point *F* and Point *M* move to Point *O*, while their terminals remain at the gridline intersections, the two halves of Arc π*D*/4 would be straightened as line *CO and JO* without altering the length of these Arcs. The proper move would create equal opposite isosceles triangles or pairs of equal diagonals of rectangles *BCJG*. Points *F* and *M* must meet at Point *O* because if the move would be short of meeting at this common point the Arcs remain bent or not flattened as straight lines. If these Arcs' midpoints passed thru Point *O* their terminals would detach from the gridline intersections at Points *CJ* and Points *B* and *G* – a condition that violates the circle component's geometric-math relationships by way of the Pythagorean Theorem. The two sides of the isosceles triangles *COJ* and *JOG* with a common apex at Point *O* are equal and the sum of which sides is also equal to a rectangle's diagonal *CG*. Hence, the straightened Arc π*D*/4 is the hypotenuse of the right triangle *CGJ*. Based upon the above analyses, the Pi value can be calculated by either equating the Arc π*D*/4 with the sum of the isosceles triangle *COJ* equal sides or with the line *CG* which is the hypotenuse of the right triangle *CGJ*.

Circle's True Pi Value Equals the Square Root of Ten

Find the true Pi value using the isosceles triangle *COJ*:

JO is the Hypotenuses of the Altitude *LO* and Base *JL*

CO is the Hypotenuses of the Altitude *LO* and Base *CL*.

Take note that non-italicized letters representing lines are used in the following equations to facilitate the math operation.

LO = D/4$\sqrt{2}$ and JL = D/2$\sqrt{2}$	Point O lies in the center of the rectangle CBGJ
(JO)² = (D/4$\sqrt{2}$)² + (D/2$\sqrt{2}$)²	Pythagorean Theorem is applied on triangle LOJ
(JO)² = (D²/32) + D²/8	Square is applied on the right side of the equation
(JO)² = D²/32 + 4D²/32	Last term is multiplied by (4/4) to get common divisor
(JO)² = 5D²/32	Right side of the equation, the terms are added
JO = $\sqrt{5D²/32}$	Square root is applied on both sides of the equation
JO = $\sqrt{5D²/(2)(16)}$	Perfect square (16) is introduced as a multiplier of 2
JO = (1/4) $\sqrt{5D²/2}$	Square root of 16 equals 4 is cleared out of the square root
CO = JO	The two sides of the isosceles with a common apex are equal
Flattened as angle shape Arc πD/4 = JO + CO	
π/4 = (2)(1/4) $\sqrt{5}$/2	Value of JO is doubled; Arc equals the sum of JO and CO
π/4 = 1/2) $\sqrt{5}$/2	D is cleared out under the square root and deleted from both sides of the equation.
π = 2 $\sqrt{5}$/2 = $\sqrt{5}$/(4)/(2)	Both sides of the equation are multiplied by 4
π = $\sqrt{(5)(2)}$ = $\sqrt{(10)}$	Square the 2 at the outside of the square root sign equals 4 and place it inside the square root sign

$\pi = \sqrt{2}\sqrt{5}$	The 5 times 2 both under the square root equals the $\sqrt{(10)}$
$\pi = \sqrt{10}$	$\sqrt{2}$ multiplied by $\sqrt{5}$ equals $\pi = \sqrt{10}$

OJ equals OG; hence CO + OJ equals CG.

Find the true value of the Pi (π) using the right triangle BCJ:

$(CG)^2 = (CJ)^2 + (GJ)^2$	Pythagorean Theorem applied on a right triangle CGJ
$(\pi D/4)^2 = (D/\sqrt{2})^2 + (D/2\sqrt{2})^2$	The values of Arc, Chord, and ½ Chord are substituted.
$(\pi D/4)^2 = (D^2/2 + (D^2/8)$	Right side squared
$(\pi D/4)^2 = (1/2 + 1/8)D^2$	Common multiplier D^2 factored out
$(\pi D/4)^2 = (4/8 + 1/8)D^2$	The term (1/2) is multiplied by (4/4).
$(\pi D/4)^2 = (5/8)D^2$	Fractions on the right side of the equation are added
$\pi^2 D^2/16 = (5/8)D^2$	Right side of the equation squared.
$\pi^2 D^2 = (80/8)D^2$	Both sides of the equation are multiplied by 16 and D^2 is canceled.
$\pi^2 = 10$	
$\pi = \sqrt{10} = 3.1623$ or 3.16227766016838	Square root is applied on both sides of the equation.

It is essential to note that the use of the old traditional $\pi = 3.14159$ will not justify the line terminals at the gridline intersections. Hence, the polygon or right triangle CGJ will not form. The following analysis further proves the validity of the right triangle formed by the Hypotenuse (straightened $\pi D/4$), the Chord, and half the Chord. The analysis uses two methods or equations in determining the areas of the triangle or the polygon. This leads to another means to derive the true Pi value. If areas from the two methods (1 and 2) would be equal, then the previous Pi derivation that the straightened arc $\pi D/4$ is equal to line CG is valid.

Circle's True Pi Value Equals the Square Root of Ten

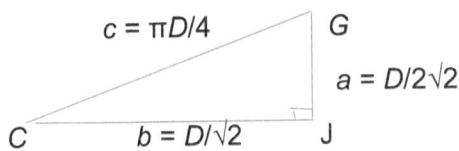

Figure 2 (Extract from Figure 1)

Method 1:
Area$_1$ = (1/2)(D/2√2)(D/√2) = D²/8   Area of right triangle or half the rectangle

Method 2:
A$_2$ = √s*(s - a)*(s - b)*(s - c) This formula is used for the polygon which is a closed plane figure consisting of at least 3 straight-line sides. In this Method 2, the terminals of sides a, b, and c should be connected end-to-end or without a gap.

Before using the formula in Method 2, it should be tested for its validity with the triangle as shown in Figure 2.
Where:

Altitude = a	Right triangle's Altitude
Base = b	Right triangle's Base
b = 2a	Right triangle's Base
Hypotenuse = c	Right triangle's Hypotenuse
Method 1: Area$_1$ = (1/2)(ab)	Area of the right triangle
Area$_1$ = (1/2)(a)(2a) = a²	
Method 2: Area$_2$ = A$_2$ = √s(s - a)(s - b)(s - c)	
s equals one-half of the sum of all sides of the polygon.	
s = (1/2)(a + b + c)	Per definition of s above
s = (1/2)(a + 2a + c)	Substitute the value of b in the equation
s = (3a/2 + c/2)	Terms in the () are multiplied by ½ and added similar terms
c² = a² + (2a)²	Pythagorean Theorem is applied
c² = a² + 4a²	Last term is squared
c = √5a	Square root is applied on both sides s = (1/2)(a + 2a + c)
s = (3a/2 + √5a/2)	Value of c is substituted in the above equation: s = (3a/2 + c/2)

13

$s = (a/2)(3 +\sqrt{5})$ — Common multiplier $(a/2)$ is introduced.

$(s - a) = (a/2)(3 +\sqrt{5}) - a$ — Side a is subtracted from s

$(s - a) = (a/2)(3 +\sqrt{5}) - a(2/2)$ — Last term a is multiplied by 2 and divided by 2

$(s - a) = (a/2)[(3 +\sqrt{5}) - 2]$ — Common multiplier $(a/2)$ is introduced

$(s - a) = (a/2)(1 +\sqrt{5})$ — 3 minus 2 equals 1

$(s - b) = (a/2)(3 +\sqrt{5}) - 2a$ — Side b equals *(2a)* is subtracted from the value of s

$(s - b) = (a/2)(3 +\sqrt{5}) - 4a/2$ — Common multiplier $(a/2)$ is introduced. Last term $2a$ is multiplied by 2 and divided by 2.

$(s - b) = (a/2)(3 +\sqrt{5} - 4)$ — Last term $(4a/2)$ is included in the brackets and $(a/2)$ is eliminated; leaving (-4) as the last term in the closed parenthesis

$(s - b) = (a/2)(\sqrt{5} - 1)$ — 3 plus (-4) equals -1

$(s - c) = (a/2)(3 +\sqrt{5}) - \sqrt{5}a$ — Side c equals *($\sqrt{5}a$)* is subtracted from the value of s

$(s - c) = (a/2)(3 +\sqrt{5} - 2\sqrt{5})$

$(s - c) = (a/2)(3 - \sqrt{5})$ — $\sqrt{5}$ plus $(-2\sqrt{5})$ equals $(-\sqrt{5})$

Substitute all values in the polygon equation:

$A_2 = \sqrt{s(s - a)(s - b)(s - c)}$

$A_2 = \sqrt{(a/2)(3 +\sqrt{5})\,(a/2)(1 +\sqrt{5})(a/2)(\sqrt{5} - 1)(a/2)(3 - \sqrt{5})}$

$A_2 = (a/2)^2\sqrt{3 +\sqrt{5})(1 +\sqrt{5})(\sqrt{5} - 1)(3 - \sqrt{5})}$ Common multiplier $(a/2)^2$ is introduced for all terms under the square root. There were four multiplier $(a/2)^2$ previously; equals $(a/2)^4$

$A_2 = (a/2)^2\sqrt{(3 +\sqrt{5})\,(3 - \sqrt{5})\,(\sqrt{5} +1)(\sqrt{5} - 1)}$ — Terms under the square root is arranged in Binomial form

$A_2 = (a/2)^2\sqrt{(3^2 - \sqrt{5}^2)(\sqrt{5}^2 - 1^2)}$ — Binomial terms further introduced

$A_2 = (a/2)^2\sqrt{(9 - 5)(5 - 1)}$ — Terms under the square root are squared

$A_2 = (a/2)^2\sqrt{(4)(4)}$ — Terms under the square roots are simplified by subtraction

Circle's True Pi Value Equals the Square Root of Ten

$A_2 = (4)(a^2)/4$ (4) is the square root of 4 times 4; the divisor 4 equals a^2

$A_2 = a^2$ 4/4 is canceled out

$A_1 = A_2 = a^2$ Hence, the formula for the area (A_2) of a polygon using all sides is valid.

The following equations prove that the straightened Arc $\pi D/4$ is the Hypotenuse of Altitude (Chord $D/2\sqrt{2}$) and Base (Chord $D/\sqrt{2}$). The analysis uses the two methods in determining the areas of the triangle or polygon which leads to another derivation of the true Pi value. If areas from the two methods (1 and 2) below would be equal, then the straightened arc $\pi D/4$ is equal to line CG.

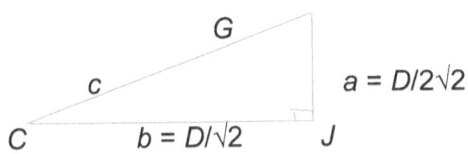

Triangle Formed in *Figure 1*

Method 1:
Area$_1$ = $(D/2\sqrt{2})(D/\sqrt{2})(1/2) = D^2/8$ Area of the right triangle

Method 2:
$A_2 = \sqrt{s*(s-a)*(s-b)*(s-c)}$ This formula is used for a polygon which is

a closed plane figure consisting of straight-line segments. Each line segment must intersect two other line segments, one at each point. In this Method 2, the terminals of sides a, b, and c should be connected end-to-end or without a gap.

If the two Pi (πs) derived from A_1 and A_2 would be equal, this means that the straightened Arc $\pi D/4$ is the hypotenuse c; that is $\pi D/4 = \sqrt{(D/2\sqrt{2})^2 + (D/\sqrt{2})^2}$ in the previous equation. The equality also would prove that the straightened Arc's both endpoints terminate at the intersections of the gridlines.

Where:

$a = D/2\sqrt{2}$ Right triangle's Altitude
$b = D/\sqrt{2}$ Right triangle's Base
$c = \pi D/4$ Right triangle's Hypothenuse
s Equals one-half of the sum of all sides of the triangle or any polygon.
$s = (1/2)*(D/2\sqrt{2} + D/\sqrt{2} + \pi D/4)$

$s = (D/4\sqrt{2} + D/2\sqrt{2} + \pi D/8)$ All terms right of equation multiplied by common (1/2)

$s = (D/4\sqrt{2} + 2D/4\sqrt{2} + \pi D/8)$ Second term multiped by (2/2)
$s = (3D/4\sqrt{2} + \pi D/8)$ First and second terms added
$(s - a) = 3D/4\sqrt{2} + \pi D/8 - D/2\sqrt{2}$ Value of a subtracted from s.
$(s - a) = 3D/4\sqrt{2} + \pi D/8 - 2D/4\sqrt{2}$ Third term multiped by (2/2)
$(s - a) = (D/4\sqrt{2} + \pi D/8)$ first and last terms added (+ – signs considered)

$(s - b) = (3D/4\sqrt{2} + \pi D/8) - D/\sqrt{2}$ Value of b subtracted from s
$(s - b) = 3D/4\sqrt{2} + \pi D/8 - 4D/4\sqrt{2}$ Last term multiplied by (4/4)
$(s - b) = (- D/4\sqrt{2} + \pi D/8)$ first and last terms added (+ – signs considered)

$(s - c) = 3D/4\sqrt{2} + \pi D/8 - \pi D/4$ Value of c subtracted from s
$(s - c) = 3D/4\sqrt{2} + \pi D/8 - 2\pi D/8$ Last term's multiplied by 2/2.
$(s - c) = (3D/4\sqrt{2} - \pi D/8)$ first and last terms added (+ – signs considered).

Substitutes all values in the polygon equation:
$A_2 = \sqrt{(3D/4\sqrt{2} + \pi D/8)*(D/4\sqrt{2} + \pi D/8)*(- D/4\sqrt{2} + \pi D/8)*(3D/4\sqrt{2} - \pi D/8)}$
Rearrange the equation in binomial term format, $(x + y)(x - y)$:
$A_2 = \sqrt{(3D/4\sqrt{2} + \pi D/8)*(3D/4\sqrt{2} - \pi D/8)*(- D/4\sqrt{2} + \pi D/8)*(- D/4\sqrt{2} + \pi D/8)}$
Use substitution to ease the computation:
Let, $x = 3D/4\sqrt{2}$, and $y = \pi D/8$

$(x + y)(x - y)$ Binomial form
$x^2 + xy - xy - y^2$ Terms are expanded
$x^2 - y^2$ Terms of equal values with opposite signs are canceled out

Let, $x_1 = \pi D/8$ and $y_1 = D/4\sqrt{2}$
$x_1^2 - y_1^2$
$A_2 = \sqrt{(x^2 - y^2)*(x_1^2 - y_1^2)}$
$A_2 = \sqrt{(3D/4\sqrt{2})2 - (\pi D/8)2)*(\pi D/8)2 - (- D/4\sqrt{2})2)}$ Values of x, y, x_1, and y_1 resubstitute

Circle's True Pi Value Equals the Square Root of Ten

$A_2 = \sqrt{(9D2/32 - \pi 2D2/64)*(\pi 2D2/64 - D2/32)}$ Terms under the square root are squared

$A_2 = \sqrt{(18D^2/64 - \pi^2 D^2/64)*(\pi^2 D^2/64 - 2D^2/64)}$ The first and the last terms multiplied by (2/2) or (2 divided by 2)

$A_2 = \sqrt{(D2/64)*[(18 - \pi 2)]*(D2/64)*[(\pi 2 - 2)]}$ Factor out the $(D^2/64)$ from both binomial terms

$A_2 = (D^2/64)*\sqrt{(18 - \pi^2)*(\pi^2 - 2)}$ Square root of two $(D^2/64)$ equals one $(D^2/64)$.

$A_2 = A_1$ Regardless of the formula usage identical polygons yield equal areas.

Because of the equality of A_1 and A_2, the π can be determined by using the Quadratic Formula: $x = (-b [+ -] \sqrt{(b^2 - 4ac)})/(2a)$.

$A_2 = (D^2/64)*\sqrt{(18 - \pi^2)*(\pi^2 - 2)}$ From the previous equation above.

$A_1 = D^2/8$ A_1 by one-half rectangle method or right triangle

$D^2/8 = (D^2/64) \sqrt{(18 - \pi^2)*(\pi^2 - 2)}$ A_1 and A_2 is equated

$(D^2/8)/ (D^2/64) = \sqrt{(18 - \pi^2)*(\pi^2 - 2)}$ Both sides of equation divided by $(D^2/64)$

$8 = \sqrt{(18 - \pi^2)*(\pi^2 - 2)}$ Left side fraction divided; $(D^2/8)/(D^2/64) = 8$

$64 = (18 - \pi^2)*(\pi^2 - 2)$ Both sides of equation are squared

$64 = 18 \pi^2 - 36 - \pi^4 + 2\pi^2$ Terms on the right side of the equation expanded

$64 = -\pi^4 + 20\pi^2 - 36$ First and last terms on the right side of the equation are added

$-\pi^4 + 20\pi^2 - 100 = 0$ Transferred 64 to the right of the equation and added to (-36)

$\pi^4 - 20\pi^2 + 100 = 0$ All terms multiplied by (-1); the same as transposition that changes all signs from positive to negative and vice versa

$(\pi^2)^2 - 20\pi^2 + 100 = 0$ First term $(\pi^2)^2$ equals π^4

Let $x = \pi^2$ Use the term substitutions

$x^2 - 20x + 100 = 0$ Quadratic equation formed after x is substituted to the equation

Find the value of x using the Quadratic Formula: $x = (-b (+ -))\sqrt{(b2 - 4ac)})/(2a)$.

Where:

$a = 1$ a is the multiplier of x^2 in the Quadratic Equation

$b = -20$ b is the multiplier of x in the Quadratic Equation

$c = 100$ c is the constant (100) in the Quadratic Equation

Substitute values of a, b, and c to the Quadratic Equation:

$x = ((-(-20) +- \sqrt{(-20)^2 - 4*(1)*(100)})/(2)(1)$ Values of a, b, and c are substituted in the Equation.

$x = (20 + \sqrt{400 - 400})/2$

$x = (20 + 0)/2$

$x = 20/2$

$x = 10$

Equate the previously substituted value of $x = \pi^2$

$x = \pi^2 = 10$

$\pi = \sqrt{10}$ Square roots applied on both sides of the equation.

Apply this Pi value to the previously derived equation that resulted from the Binomial Equation: $A_2 = (D^2/64)*\sqrt{(18 - \pi^2)*(\pi^2 - 2)}$

$A_2 = (D^2/64)*\sqrt{(18 - \sqrt{10}^2)*(\sqrt{10}^2 - 2)}$ New Pi value is substituted.

$A_2 = (D^2/64)*\sqrt{(18 - 10)*(10 - 2)}$ The square and square root are applied on 10.

$A_2 = (D^2/64)*\sqrt{(8)*(8)}$ Substructions under the square root are performed.

$A_2 = (D^2/64)*(8)$

$A_2 = D^2/8$

Also, $A_1 = D^2/8$

Because A_1 equals A_2, the two triangles are congruent and the straightened Arc $\pi D/4$ is equal to the line CG. Hence, the previous equations used in deriving the true Pi ($\pi = \sqrt{10} = 3.1622$) value are valid. The equation for the area of a closed figure (polygon) that derived the true value of Pi validated that the straightened value of the Arc, $\pi D/4$ is equal to $\sqrt{(D/2\sqrt{2})^2 + (D/2\sqrt{2})^2}$.

Besides, Triangle 1 and Triangle 2 are congruent because they have equal areas and identical Altitudes and Bases; hence, the straightened Arc πD/4 (line CG) is the hypotenuse of half the Chord and a full Chord of a Circle formed by a 90 degrees central angle. The Pythagorean Theorem applies to Triangle 2: $(πD/4)^2 = (D/2\sqrt{2})^2 + (D/\sqrt{2})^2$ which yielded a Pi (π) value = $\sqrt{10}$ = 3.1622 or 3.16227766016838 per previously calculated in this article. In other words, the congruency of Triangle 1 and Triangle 2 proved that the previously calculated π = $\sqrt{10}$ is precise and valid. The resulting Pi (π) value = $\sqrt{10}$ = 3.1622 or 3.16227766016838.

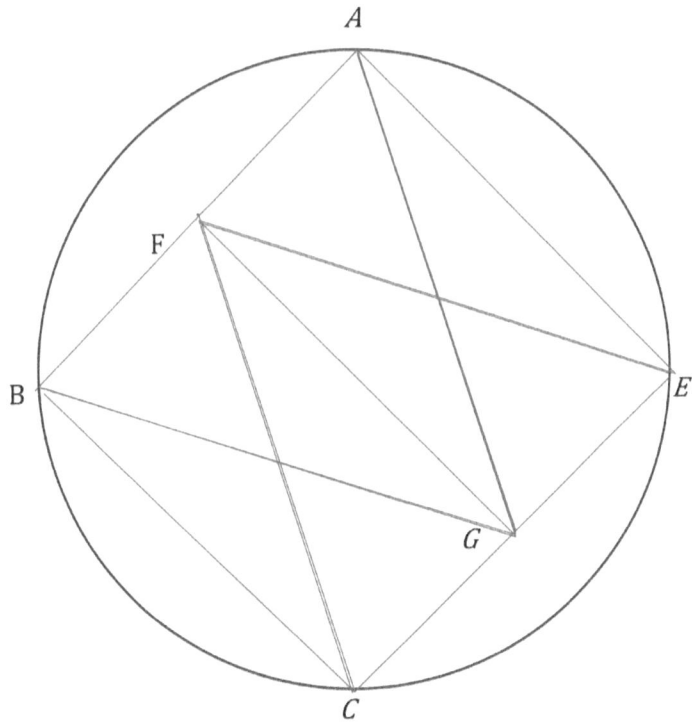

Figure 3

In *Figure 3*, the Points and Lines descriptions are as follows:

Points A, B, C, and E lie on the Circle's circumference.
Point D is not used in the figure as it is reserved for the diameter.

Point F bisects line AB	Line FG divides the square into two equal parts.
Point G bisects line CE	Line FG divides the square into two equal parts.
AB = BC = CE = AE	Square's sides are equal.
AEGF = BCGF	Two identical rectangles formed by the inscribed squares are equal.
AF = EG = BF = CG	Two sets of identical rectangles' opposite widths are equal.
AG = EF = BG = CF	Two sets of identical rectangles' diagonals are equal.
BC = 2CG	BC equals the square's side while CG is half the side.

The straightened Arc BC ($\pi D/4$) equals line BG as illustrated previously in this book.

In *Figure 1* and *Figure 3*, all components of the circle's end lines terminate either at the intersections of the gridlines or at the line intersections of their sub-component's parts as in the case of the rectangle's diagonals. Without such conditions, the relationships among the circle's components and sub-components would not exist. Besides, the Pythagorean Theorem would not apply unless the polygon is formed like that of the right triangles in the two figures.

In other words, if the relationships among the circle's components and sub-components do not exist, *Figure 1* and

Figure 3 are not possible. However, these figures are derived from mathematics procedures and geometric logic.

Given the scenario, only the square root of ten fits into a true Pi value. If the Pi value is slightly less or slightly greater than the square root of ten, the true relationships among the circle's components and sub-components can not be computed. Hence, Pi equals the square root of ten.

Figure 3 with its description and analysis yields a new theory: "The circle's circumference equals the two sets of diagonals of the two identical rectangles formed by the inscribed square."

CHAPTER 3
— LITERATURE REVIEW —

Traditional Method 1. Manual Measurement.

Manual measurement of the circumference of the circle and its diameter is not a precise method. Any measuring device has a thickness whose centerline lies along the middle of the thickness throughout its entire length. Wrapping the string or measuring around the circle is equivalent to taking the circumference of the circle with a wider diameter (i.e., the actual diameter plus the thickness of the string or rope). *Figure 3* illustrates this situation. The error exacerbates with the use of a thicker measuring device. Hence, the traditional Pi π = 3.1415 is slightly lower than its true value.

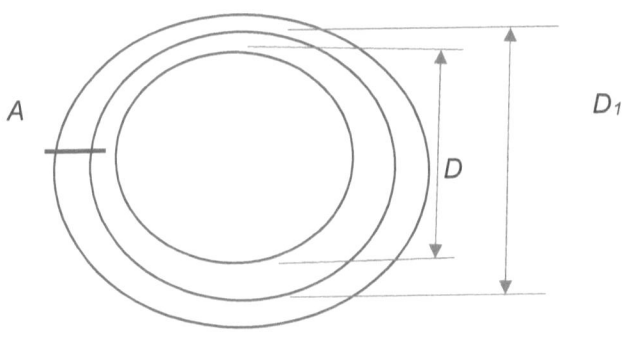

Figure 3

The thickness of the rope is exaggerated here to enhance the illustration. Where: D is the diameter of the circle; D_1

is the false diameter resulting in wrapping the rope. D_1 measures from the center to the center of the opposite thickness of the rope. Point A is the terminal of the rope around the circumference (πD_1). Since D1 is slightly longer than D, the traditional calculated Pi is slightly less than the real value of the Pi. Besides, the problem with this method is the inaccuracy in arriving at four or more digits from the decimal places of 3.14.

Instead of manual measurements, the contemporary author of this new book used a different approach to determining the true Pi value. The contemporary author used a flexible wire to validate the equality of the Arc $\pi D/4$ and line CG. A straight wire was cut equal to the length of line CG. The wire was smoothly bent until it was shaped like the Arc CFJ ($\pi D/4$). Without altering the wire's length, the curved wire fitted congruent with the Arc CFJ ($\pi D/4$). The experiment was repeated many times using new wires at a time and the results were consistent. This experiment confirmed that the Arch CFJ ($\pi D/4$) is equal to line CG which is the hypotenuse of the Base chord and the Altitude half of the chord. The thickness of the wire in this experiment is irrelevant because it is not wrapped around the circle which causes an error in taking the ratio between the circumference and the diameter.

Traditional Method 2. The use of polygons to approximate the Pi (π)

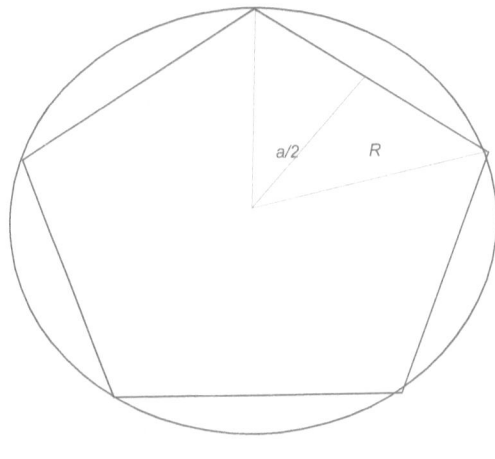

Figure 4

The Ancient Greek mathematician Archimedes discovered an effective method for approximating the value of Pi (π). With the use of a similar polygon as in *Figure 4*, he inscribed a regular hexagon in a circle and then circumscribed another regular hexagon in the same circle. He obtained a rough approximation of Pi (π) by dividing the circumferences and diameters of the hexagons. Archimedes used a 96-side polygon, which helped him find the closest approximation of Pi (π) as the original straight sides began to shape into a circle. The value of 3.14 has remained one of the most prevalent approximations of Pi (π) ever since.

Circa 600 years after the initial introduction of Pi approximation, the Chinese mathematician Zu Chongzhi utilized a similar method with a 12,288-side polygon, which yielded 6 decimal places. Nearly another 600 after Zu a new method was introduced with the use of the equation:

Perimeter $nL = n \times \sin(180°/n)$. The value of this traditional Pi is slightly less than the value of the true Pi ($\pi = \sqrt{10} = 3.1622$) because, for the $n \times L$ to equal the circumference πD, $180°/n$ must equal zero, which requires n to reach infinity ∞.

Traditional Method 3. Calculating Pi (π) using infinite series

Mathematicians eventually introduced a near-exact formula for calculating Pi (π); however, this requires infinite complex calculations. One of the most well-known methods to calculate Pi (π) is to use the Gregory-Leibniz Series: $\pi=4(1-13+15-17+...)\pi=4(1-13+15-17+...)$

This proof starts with the Taylor series: $11-y=1+y+y2+...11-y=1+y+y2+...$ Apply the variable substitution $y=-x2y=-x2$ to get: $11+x2=1-x2+x4-x6+...11+x2=1-x2+x4-x6+...$

Traditional Method 4.
Another series that unites more quickly is the Nilakantha Series. During the 15th century, Nilakantha developed a series that requires fewer terms to derive a much closer value to the Pi π. It is somewhat similar to the previous method and also one of the conventional methods. As the number of iterations increases the value of Pi also gets precise.

$$\pi = 3 + \frac{4}{2*3*4} - \frac{4}{4*5*6} + \frac{4}{6*7*8} - \frac{4}{8*9*10}...$$

Contemporary mathematicians have also found other more efficient series for calculating Pi (π). Computer programs can add up more and more terms, calculating Pi (π) to extraordinary closer degrees of accuracy. In 2014 the world record was that a computer has calculated Pi (π) correct to 13,300,000,000,000 decimal places. Before the advent of computers, it was much harder to calculate Pi (π). In the 19th Century, William Shanks took 15 years to calculate Pi (π) correctly to 707 decimal places. Unfortunately, it was later found that he had made a mistake and was only right to 527 decimal places! The 9 or ten digits of Pi (π) have been available on calculators for a century. The problem with these methods was that the value of Pi remains the same after the nth number of decimals.

Traditional Method 5. Ramanujan's Pi Formula

Ramanujan's Pi formula is one of the best methods to find a numerical approximation of Pi in less number of iterations. It may look difficult to implement but that is not the case, it's pretty simple, just follow these steps.

$$\frac{1}{\pi} = \frac{2\sqrt{2}}{99^2} \sum_{k=0}^{\infty} \frac{(4k)!}{k!^4} \frac{26390k + 1103}{396^{4k}}$$

Though the Time Complexity is higher than in previous approaches, in this approach, one will need significantly less number of iterations so this is considered to be an effective technique.

Method 5. Calculating Pi using Elementary Calculus

There are many powerful and sophisticated methods known for the calculation of Pi, and, using these methods, Pi has been calculated to billions of decimal places. Here we present two methods for the calculation of Pi which only use elementary calculus, but are surprisingly effective. By way of comparison, note that *Pi = 3.14159265358979323846 and so on.*

Traditional Method 6. The Use of Limits

This method uses the fact that

$$\lim_{x \to 0} \frac{\sin(x)}{x} = 1$$

But $x = Pi/n$ for n large, Pi is approximately equal to $n \cdot \sin(Pi/n)$.

To use this, we must be able to compute $\sin(Pi/n)$. The trigonometric identities show the following relationships:

$\sin(x/2) = sqrt((1-\cos(x))/2)$, $\cos(x/2) = sqrt((1+\cos(x))/2)$,

valid for $0 <= x <= Pi$, where we take the positive square roots. Thus, if we know $\sin(Pi/n)$ and $\cos(Pi/n)$, we may compute $\sin(Pi/2n)$ and $\cos(Pi/2n)$, and then proceed recursively. Of course, we know $\sin(Pi/6) = 1/2$ and $\cos(Pi/6) = sqrt(3)/2$.

Traditional Method 7.

Beginning with the integral

$$\tan^{-1}(x) = \int_0^x \frac{dt}{1+t^2}$$

and writing $1/(1 + t^2) = 1 - t^2 + t^4 - t^6 + ...$ and integrating term-by-term, we obtain the infinite series

$$tan^{-1}(x) = x - x^3/3 + x^5/5 - x^7/7 + ...,$$

which converges for -1 <= x <= 1. Substituting $x = 1$, we obtain Leibniz's series

$$Pi = 4(1 - 1/3 + 1/5 - 1/7 + 1/9 - ...)$$

However, because $x = 1$ is an endpoint of the interval of convergence, this series converges very slowly and so is impractical for calculation. (For example, the first five terms give the approximations 4., 2.666, 3.466, 2.895, 3.339 to Pi.)

On the other hand, we may substitute $x = 1/sqrt(3) = tan^{-1}(Pi/6)$. Then, after some algebraic manipulation, we obtain the series

$$Pi = 2 \cdot sqrt(3)(1 - 1/3 \cdot 3 + 1/5 \cdot 3^2 - 1/7 \cdot 3^3 + 1/9 \cdot 3^4.)$$

and the first five terms of this series give the approximations 3.46410, 3.07920, 3.15618, 3.13785, and 3.14260 to Pi, the last of which differs from Pi by little more than .001.

Indeed, we can do even better. Substituting $x = tan(Pi/n)$ for $n >= 4$ (so that the series converges), we obtain the series

$Pi = n(tan(Pi/n) - tan3(Pi/n)/3 + tan5(Pi/n)/5 - tan7(Pi/n)/7 + tan9(Pi/n)/9 - ...)$.

Now $tan(x/2) = sqrt((1 - cos(x))/(1 + cos(x)))$, so we may again compute $tan(Pi/n)$ recursively, starting with $n = 6$. We obtain the following approximations to *Pi*, where in the last two columns of this table, $S_k(tan(Pi/n))$ denotes the approximation obtained by using the first *k* terms of the above series.

The existence of a formula that is faster than calculating the nth binary digit of π remains an open question. The ever-increasing number of decimal places onto the original Pi (π) = 3.1415 is merely a continual use of the erroneous initial value of the most important number in the world. It is also referred to as Archimedes's constant. Contemporary mathematicians have also found other more efficient series for calculating Pi (π). Computer programs can add up more and more terms, calculating Pi (π) to extraordinarily closer degrees of accuracy. In 2014 the world record was that a computer calculated Pi (π) correctly to 13,300,000,000,000 decimal places.

The following paragraphs explore some of the global highlights of Pi. This history includes some mathematical concepts of limits and infinite series.

There is no single origin of the Pi number. The Greek letter π is used due to the influence of its origins in Greece. The Pi simply represents the ratio between the circumference and diameter of a circle. Ever since mathematicians have been thinking about circles and discovering new ways to approximate and calculate π. The number π has been studied, approximated, and thought out since circa 3000 before the Christian era (BCE). Many mathematicians and thinkers helped advance the knowledge of numbers.

Circa (Ca.) 3000 BCE: The first known people to search for π during this era were Babyloniansinas and EgyotiansThe Egyptian pyramids of Cheops and Sneferu at Gizeh both have a ratio of half the perimeter to the height equal to 3.17. Perhaps, this ratio is an early attempt at calculating the ratio between the perimeter of a circle and its diameter.

1850 BCE: An Egyptian named Ahmes scribed in Rhind Papyrus the early attempt to document Pi. He calculated the surface area of a hemisphere involving circles with the use of $π = (16/9)^2 = 3.1605$. Though this number was not accurate, it served as a basic number for calculating circles and alike areas for many years during those times.

Ca. 1500 BCE: The Greek mathematician Archimedes first used mathematics to compute π which illustrated a value between 22/7 and 223/71. If the Diameter is seven Inches in length, then the Circumference is 22 Inches. He proceeded by inscribing a polygon in a circle and circumscribing another polygon in that circle. The perimeter of the inner polygon is less than the circle's circumference while the perimeter of the outer polygon is greater than that of the circle. Archimedes took the average of the two polygons' perimeters. Using a hexagon, a 12-sided polygon, a 24-sided polygon, a 48-sided polygon, and a 96-sided polygon Archimedes discovered that all of them yielded a range between 3.1408 and 3.1428.

Ca. 440 BCE: The Greek mathematician Antiphon took the revolutionary step of calculating Pi. He inscribed in a circle a polygon of an ever-increasing number of sides, from which he discovered the limits of one of the principles of calculus. The calculation of the area of a circle would reveal the value of π because the area of a circle with radius r is given by $A=\pi r^2$.

ca. 265: A Chinese mathematician Liu Hui independently discovered a similar approach as Antiphon. Lui used a thousand-sides polygon. He determined the first four digits after the decimal point of π (3.1415). Around 200 years later, the Chinese mathematician Zu Chongzhi calculated π using Liu Hui's algorithm applied to a 12,288-sided polygon, the most accurate approximation for nearly a millennium (3.141592920...).

ca. 1360 AD: Historically, the first thought exact formula for π used infinite series and was not available until around 1400. Medieval Indian mathematician-astronomer Madhava discovered the series, whose discovery remained unknown in the West until relatively recently. Though almost all of Madhava's original work is lost, he is referenced often in later mathematical works and represents early steps away from the finite processes of algebra into considerations of the infinite. He discovered that one can calculate π using the following infinite series, now known as the Madhava-Leibniz or Madhava-Gregory-Leibniz series, crediting other mathematicians who independently discovered the series centuries later: π4=1–13+15–17+19–⋯.

Ca. 1424 AD: a Persian astronomer and mathematician Jamshid al-Kashi calculated π using a polygon with 3,228 sides. Al-Kashi generated a number able to calculate the size of the universe in effect setting the world record for 180 years. While it is true that the decimal representation of π has an infinite number of digits, in truth, modern-day NASA would only need around 16 digits of π to be able to calculate precise distances for orbiting spacecraft—a benchmark achieved centuries before NASA even existed!

In 2002: Discovering more digits of π is no longer very significant to mathematics. However, it has more meaning to computer scientists to showcase human ingenuity. Japanese mathematician Yasumasa Kanada set multiple records for computing π with over 1 trillion decimal places.

In 2019: A Japanese computer scientist Emma Haruka Iwao set a world record when Iwao and her team calculated over 31.4 trillion digits of π.

In 2021: Researchers from the University of Applied Sciences of the Grisons in Switzerland used a supercomputer that runs for 108 days which resulted in a mind-boggling 62.8 trillion decimal digits in Pi.

CHAPTER 4
RESULTS

Based on this new geometric math analysis, the Pi value ($\pi = \sqrt{10}$ = 3.1622 or 3.16227766016838), truly represents the precise ratio between the Circumference and the Diameter of a Circle. Because the Pythagorean Theorem rules in any relationships among the components of the circle, the Pi value should be the square root of a real number. The traditional methods of calculating Pi by many scientists and mathematicians are merely approximations or short of representing the precise value of the Pi. The old calculations failed to produce the true ratio between the circumference and the diameter of a circle. The traditional Pi value of 3.1415 is not a perfect square root of a real number. The key to solving the mystery of the true value of Pi is to establish the relationships between the arc of the circle and the resulting chord. This innovative approach to geometric math derivation of the Pi number stops at the 14th decimal place because of its accuracy. This contemporary geometric derivation of Pi ($\pi = \sqrt{10}$) refutes the previous methods of finding the value of Pi based on the predecessors' works that yielded π = 3.14159265359. Any effort to introduce more digits to this constant is futile because the base value is already inaccurate.

CHAPTER 5
SUMMARY

The proof that Pi equals the square root of ten begins with the construction of *Figure 1* consisting of a circle inscribed by a square with an equal diameter (D) and side (S). The square is divided into grids of D/4 wide for each grid. Using a central angle of 90 degrees, the set of radius (D/2) forms an Arc equal to πD/4 and 4 sets of equal D/√2 Chords. Because all the circle's major components terminate at the grid intersection, it was assumed that the equivalent straightened Arc is also the same as the other components. One end of the Arc stayed in place while the other end moved to the nearest intersection. This movement created a right triangle whose two perpendicular sides are one full chord and one-half chord, and the straightened Arc is the hypotenuse. The extension of the Chord (*D/√2*) into a Secant of a circle serves as one of the equal sides of the Isosceles triangle GMC (*Figure 2*) which component lines and angles are analyzed with the use of the Tangent Formula, Sine Law, and Pythagorean Theorem. Besides, the two methods of finding the areas of a triangle proved the validity of the assumption that the straightened arc is the hypotenuse of a right triangle. The two methods of finding the areas of a closed plane figure: the product of the base and altitude of a right triangle and the use of the formula for an irregular polygon (unequal sides closed plane figure); that is, the area is the square root of [s(s - a)(s - b)(s - c)]. Where s equals one-half of the sum of all sides of the polygon a, b, and c. The equality of the areas and the congruency of the two polygons presented by two methods proved that the straightened arc is the hypotenuse of a triangle and that all three sides are connected end-to-end without a gap. Hence, the assumption and equations used in deriving the Pi ($\pi = \sqrt{10} = 3.1622$) are valid.

CHAPTER 6
— RECOMMENDATION —

The current value of old traditional Pi (π = 3.1415 should be replaced by the square root of 10; that is $\sqrt{10}$ = 3.1623 or 3.16622776616838 as the new Pi number. Granted, most scientific applications don't need π beyond a few hundred digits, but that isn't stopping anyone; starting in 2009, engineers have used customized personal computers to calculate trillions of digits of π. The race to calculate more π digits has only accelerated as of late, with computer scientists using it as a way to test supercomputers, and mathematicians competing against one another. The algorithm for computing the Pi is too complex. In layman's terms, this means that the time and resources necessary to calculate digits increase more rapidly than the digits themselves. Furthermore, it gets harder to survive a potential hardware outage or failure as the computation goes on. The quest of finding more decimal digits should have had not happened if the original calculation of the traditional PI had been accurate.

The change of the Pi value from 3.1415 to 3.1623 is necessary for the accuracy of calculations throughout the universe, however, this would require reprogramming of computers that are currently set with the traditional value of Pi.

This correction is essential for elementary and scientific applications, especially on large-scale usages such as in astronomy and cosmology. The use of a true and precise value of Pi is also important because its most elementary definition relates to the circle, π is found in many formulae in trigonometry and geometry, especially those concerning circles, ellipses, and spheres. Besides, the change of value will put to rest the never-ending quest to introduce more digits to the traditional Pi based on 3.1415, which in the year 2021 at records of 31.4 trillion π digits with neither benefit nor impact on the actual scientific calculation process. The world of math and sciences is full of Pi records just waiting to be broken; however, this may not be necessary with the introduction of the new true Pi value, the square root of 10.

REFERENCE

Apostol, Tom (1967). *Calculus, volume 1 (2nd ed.). Wiley*, p. 102. https://en.wikipedia.org/wiki/Pi - cite_ref-49https://en.wikipedia.org/wiki/Pi - cite_ref-14

Area of Unequal Sides Polygon. Downloaded from https://www.civilconcept.com. Retrieved May 12, 2022

Bogart, Steven. *"What Is Pi, and How Did It Originate?" Scientific American*. Retrieved 10 August 2020.

Borwein, J.M.; Borwein, P.B.; Dilcher, K. (1989). "Pi, Euler Numbers, and Asymptotic Expansions". *American Mathematical Monthly*. 96 (8): 681687. doi:10.2307/2324715. hdl:1959.13/1043679. JSTOR 2324715. https://en.wikipedia.org/wiki/Pi - cite_ref-96

Computation of Pi by Archimedes: *File Exchange, MathLab Central. Mathworks.com*. Archived from the original on February 25, 2013. Retrieved March 12, 2013.

Euler, Leonhard (1747). Henry, Charles (ed.). *Lettres inédites d'Euler à d'Alembert. Bullettino di Bibliografia e di Storia delle Scienze Matematiche e Fisiche* (in French). Vol. 19

(published 1886). p. 139. E858. https://en.wikipedia.org/wiki/Pi - cite_ref-117

Eymard, Pierre; Lafon, Jean Pierre (1999). The Number Pi. *American Mathematical Society. ISBN 978-0-8218-3246-2.,* English translation by Stephen Wilson.

https://en.wikipedia.org/wiki/Pi - cite_ref-hayes-2014_93-0 Hayes, Brian (September 2014). "Pencil, Paper, and Pi". *American Scientist. Vol. 102, no. 5. p. 342. doi:10.1511/2014.110.342. Retrieved 22 January 2022.*

Borwein, J.M.; Borwein, P.B. (1988). "Ramanujan and Pi". *Scientific American. 256 (2): 112– 117. Bibcode:1988SciAm.258b.112B. doi:10.1038/Scientific American 0288- 112.*

Haruka Iwao, Emma (14 March 2019). "Pi in the sky: Calculating a record-breaking 1.4 trillion digits of Archimedes' constant on Google Cloud". *Google Cloud Platform. Archived from the original on 19 October 2019.* Retrieved 12 April 2019.

Jones, William (1706). *A new introduction to Mathematics. pp 243, 263. Archived from the original on 25 March 2012.* Retrieved 15 October 2021.

Milla, Lorenz (2018), A detailed proof of the Chudnovsky formula with means of basic complex analysis, arXiv:1809.00533

https://en.wikipedia.org/wiki/Pi - cite_ref-207 Peitgen, Heinz-Otto, *Chaos, fractals: new frontiers of science*, Springer, 2004, pp. 801–803, ISBN 978-0-387-20229-7.

https://en.wikipedia.org/wiki/Pi - cite_ref-12 "Pi". Dictionary.reference.com. 2 March 1993. Archived from the original on 28 July 2014. Retrieved 18 June 2012.

"Pi (π) trillion digits of π". Pi2e.ch. Archived from the original on 6 December 2016.

https://en.wikipedia.org/wiki/Pi - cite_ref-231 Rosenthal, Jeffrey S. (February 2015). "Pi Instant". Math Horizons. 22 (3):22. doi:10.4169/mathhorizons.22.3.22. S2CID 21854-2599.

https://en.wikipedia.org/wiki/Pi - cite_ref-114 Schepler 1950, p. 220: William Oughtred used the letter π to represent the periphery (that is, the circumference) of a circle.

https://en.wikipedia.org/wiki/Pi - cite_ref-115 Segner, Joannes Andreas (1756). *Mathematic Course*. Halae Magdeburgicae. p. 282. Retrieved 15 October 2020.

Shanks, William (1853). Contributions to Mathematics: Comprising Chiefly the Rectification of the Circle to 607 Places of Decimals. Macmillan Publishers. p. viii – via the Internet Archive.

Shanks, William (1873). On the extension of the numerical value of π. Royal Society Publishing. p. 318–319. doi:10.1098/rspl.1872.0066.

APPENDIX A

Glossary

Adjacent angles - Two angles that have a common vertex and common side.

Altitude - A perpendicular segment from a vertex to its opposite side.

Angle - A measure formed by two rays that have the same endpoint.

Angle bisector - A ray that separates an angle into two congruent angles. In a triangle, a segment is drawn from the vertex and cuts the vertex angle in half.

Arc - A portion of a circle's circumference.

Area - The number of square units covering a two-dimensional figure.

Base - The side or face to which an altitude can be drawn. Or the bottom side of a right triangle.
Center of Rotation - The fixed point about which a figure is rotated. Or the point about which a figure is rotated.

Central angle - An angle whose vertex is located at the center of a circle.

Chord - A segment that connects two points on a circle.

Circle - A figure that has an endless number of points equidistant from a given point. Or a round plane figure whose boundary (the circumference) consists of points equidistant from a fixed point (the center).

Circumference - The distance around the outside of a circle.

Clockwise - To rotate in the direction of a clock's hands.

Conclusion - The part after "then" in a conditional statement.

Concurrent lines - Three or more lines intersect at the same point.

Congruent - Figures that are the same shape and size. Or figures whose corresponding side's length and corresponding angles are equal to each other.

Cosine - A trigonometric ratio between the adjacent side and the hypotenuse of a triangle.

Counterclockwise - To rotate in the opposite direction of a clock's hands.

Cylinder - A solid figure with two parallel, congruent circular bases.

Deductive Reasoning - The process of using facts, definitions rules, and properties or form a logical argument.

Degrees - A unit of measure used for angles.

Diagonal - A line segment connecting two non-consecutive vertices.

Diameter - A segment that connects two points on a circle and passes through the center.

Endpoints - Points showing where a line segment begins and ends

Equiangular polygon - A polygon with angles of equal measure.

Equiangular triangle - A triangle with three congruent angles.

Fixed point - The point about which a figure is rotated.

Geometry - A branch of mathematics that deals with the measurement, properties, and relationships of points, lines, angles, and surfaces.

Given - The information provided about a problem that can be used in the proof.

Hypotenuse - The side across from a right angle in a right triangle.

Initial point - The starting point of a vector

Interior angle - Each angle is enclosed in a polygon.

Intersecting lines - Lines that cross or meet.

Irregular polygon - A polygon that is neither equilateral nor equiangular.

Isosceles triangle - A triangle whose adjacent sides with common vertex are equal.

Legs - The adjacent sides of the right angle in a right triangle.

Lines - A straight, one-dimensional figure that extends infinitely in both directions.

Major arc - An arc measures greater than 180 degrees.

Minor arc - An arc measures less than 180 degrees.

n-gon - A polygon with n sides.

Parallel lines - Coplanar lines that do not intersect.

Perimeter - The distance around a two-dimensional figure.

Pi - The number π is a mathematical constant that is the ratio of a circle's circumference to its diameter, which by tradition it is approximately equal to 3.14159.
Poin - An exact location, usually represented by a dot.
Point of tangency - The point where a tangent intersects a circle.

Polygon - A closed-plane figure consisting of straight line segments, in which each line intersects two other line segments.

Postulates - Statements that are expected without proof.

Proportion - An equation written as two ratios equal to each other.

Quadrilateral - A polygon with four sides.

Radius - A segment that connects the center of a circle to a point on the circle.

Ratio - The comparison of two quantities.

Rectangle - A parallelogram with four right angles.

Right angle - An angle that measures exactly 90 degrees.

Right triangle - A triangle with one right angle.

Rotation - A congruence transformation that turns an image about a point.

Secant - A line that intersects a circle at two points.

Sector - A portion of a circle enclosed by two radii and the intercepted arc.

Segment - Part of a line that does not extend infinitely in both directions, but it has two endpoints.

Semicircle - An arc that has a 180-degree central angle.

Similar figures - Figures whose corresponding side lengths are proportional and whose corresponding angles are congruent.

Similar triangles - Triangles with congruent angles and proportional side lengths.

Sine - A trigonometric ratio between the opposite side and the hypotenuse of a triangle.

Square - A regular quadrilateral with four equal sides and four equal angles.

Supplementary angles - Two angles whose measures add up to exactly 180 degrees.

Tangent - A trigonometric ratio between the opposite side and the adjacent side of a right triangle.

Tangent line - A line intersects a circle at exactly one point.

Terminal point - The ending point of a vector.

Theorem - A true statement that can be proven.

Vector - A quantity that represents magnitude and direction.

Vertex - The common endpoint of the two lines, rays, or segments that form an angle.

Vertical angles - Two angles are formed by intersecting lines that do not share a common side and have the same measure.

APPENDIX B

Notary Public Document

AFFIDAVIT
THE STATE OF CALIFORNIA

COUNTY OF CONTRA COSTA

KNOW ALL PERSONS BY THIS PRESENCE:

I, Albert Vitales Cruz, United States Citizen, of legal age, single widower, with residence and postal address at 2651 Lucas Avenue, Pinole, California 94564. I am mentally competent to sign this Affidavit with the following statements.

1. That I have solely discovered the precise value of a circle's Pi

(π) equals 3.1622; up to the 14th decimal place) based on geometric and mathematical equations. Pi is the ratio between the circumference of a circle and its diameter.

2. That I have completed a draft of the scientific proof for my discovery which I intend to publish for public use, especially for science and academic applications.

3. That my newly discovered Pi value is intended to correct the current traditional Pi value (π = 3.1415) of which sources were all based on mathematical approximation. This traditional value motivated a vast number of mathematicians who attempted to find the nearest approximation for the true Pi value, but to my knowledge, no one has ever introduced and or published any precise method.

All efforts were focused on introducing more decimal places to the erroneous old base number.

4. And that I prepared and signed this declaration of facts to attest to the truth of my discovery and to reserve all the rights as the originator.

FURTHER AFFIANT SAYETH NAUGHT

AFFIANT   APRIL 11, 2022
                                                       Date Signed

CALIFORNIA ACKNOWLEDGMENT   CIVIL CODE S 1189

A notary public or other officer completing this certificate verifies only the identity of the individual who signed the document to which this certificate is attached, and not the truthfulness, accuracy, or validity of that document.

State of California
County of __Contra Costa__ }

On __April 11, 2022__ before me, __CORINA DIGRAZIA-NOTARY PUBLIC__
      Date   Here Insert Name and Title of the Officer
personally appeared __Albert V. Cruz__
                                                  Name(s) of Signer(s)

County of

who proved to me on the basis of satisfactory evidence to be the person(s) whose name(s) is/are subscribed to the within instrument and acknowledged to me that he/she/they executed the same in his/her/their authorized capacity(ies), and that by his/her/their signature(s) on the instrument the person(s), or the entity upon behalf of which the person(s) acted, executed the instrument.

I certify under PENALTY OF PERJURY under the laws of the State of California that the foregoing paragraph is true and correct.

WITNESS my hand and official seal.

CORINA DIGRAZIA
Notary PubUc - California
Contra Costa County
Commission # 2358915
MY Comm. Expires May 25. 2025

Signature _____
                    Signature of Notary Public
Place Notary Seal and/or Stamp Above

OPTIONAL
Completing this information can deter alteration of the document or fraudulent reattachment of this form to an unintended document.

Description of Attached Document
Title or Type of Document: _____
Document Date: _____ Number of Pages: _____
Signer(s) Other Than Named Above: _____

Capacity(ies) Claimed by Signer(s)
Signer's Name: _____ Signer's Name: _____
O Corporate Officer — Title(s): _____ O Corporate Officer — Title(s): _____
O Partner — O Limited o General   o Partner — o Limited o General
O Individual O Attorney in Fact O Individual o Attorney in Fact O Trustee o Guardian or Conservator Trustee o Guardian or Conservator O Other: _____ o Other: _____
Signer is Representing: _____ Signer is Representing: _____

02018 National Notary Association

www.ingramcontent.com/pod-product-compliance
Lightning Source LLC
Chambersburg PA
CBHW031544210526
45464CB00003B/1146